Nitra The Elemental Dragon With Nitrogen Magic

Element 7 – Magical Elementals Chemistry Fantasy

By Sybrina Durant

Nitra The Elemental Dragon With Nitrogen Magic

Element 7 – Magical Elementals Chemistry Fantasy

Paperback ISBN 978-1-942740-76-6

BISAC Codes:
JUV036000 - JUVENILE FICTION / Science & Technology
JUV037000 - JUVENILE FICTION / Science Fiction, Fantasy, Magic
JUV002270 JUVENILE FICTION / Fantasy / Dragons, Unicorns & Mythical Creatures

League City, Texas, United States of America

Contact Sybrina@sybrina.com.

NO METAL
NITRA
The Dragon with
Nitrogen Tipped Tail
NO MAGIC

Chapter 1: The Frozen Messenger

The sky above the Cobalt Reaches shimmered with the particular blue-white luminescence that heralded dawn in the world of MarBryn — a vast, extraordinary realm where ancient alchemy walked hand-in-hand with crystalline technology, and where every civilization whispered the sacred philosophy like a prayer: "No Metal, No Magic."

Nitra stretched her great wings against the morning air, their feathered ridges catching the light like the faceted edges of sapphire. She was breathtaking in the way that natural phenomena are breathtaking — effortlessly, without trying. Her scales shifted from deep cobalt along her powerful flanks to a pale, almost luminous silver-white beneath her sinuous neck. Her angular head, crowned with elegant ivory horns that curved like crescent moons, turned slowly as she surveyed the Reaches from her clifftop perch above the city of Verenholm.

Verenholm was a marvel. Built into the basalt cliffs of the northern coast, it was a city of both antiquity and innovation. Ancient nitrogen-stone towers — carved from rock infused with elemental salts millennia ago — rose alongside modern atmospheric processors that harvested nitrate compounds from the sea air to power everything from luminescent streetlamps to mechanized irrigation channels. The citizens of Verenholm understood better than most that nitrogen, Element 7 on the great Periodic

Codex, was the invisible backbone of their entire world.

Nitra understood it most of all.

She exhaled slowly, her breath misting in the cold morning air, and descended from the cliff's edge in one fluid motion, her wings spreading wide as she rode the nitrogen-rich coastal currents upward in a graceful spiral. Flying was thinking for Nitra. The altitude gave her perspective — on problems, on decisions, on the world below.

"You're going to freeze my breakfast again if you keep doing that breathing thing."

The voice came from below. Nitra banked and descended with effortless precision, landing beside a figure seated on a flat basalt boulder with a tin pan balanced precariously over a small chemical heating coil.

Brann Cendrel was small for a human, wiry and sun-darkened, with clever amber eyes and ink-stained fingers that seemed permanently tattooed by his trade. He was an alchemical cartographer — one of the finest in MarBryn — a man who could read elemental signatures in soil and translate them into living maps that pulsed with color depending on mineral concentration. He wore a dozen leather satchels of varying sizes and a pair of brass-framed magnification spectacles pushed up on his forehead. He was twenty-nine years old, perpetually cheerful, and possessed absolutely no magical ability whatsoever.

What he did possess was an encyclopedic knowledge of MarBryn's geography and the resourcefulness of someone who had survived every bad situation it could throw at him.

"I was not going to freeze your breakfast," Nitra said calmly, folding her wings and settling beside him. "I was simply breathing."

"Nitrogen-dragon breathing," Brann replied, poking his eggs suspiciously. "There's a difference."

"You are dramatic."

"I am accurate." He grinned and handed her a rolled parchment. "This arrived before dawn. Courier pigeon, but not an ordinary one. Look at the seal."

Nitra took the parchment delicately between two curved claws and examined the seal pressed into deep violet wax. It was the emblem of the Nitrate Council — seven interlocking circles representing the seven most important elemental orders of MarBryn — but the wax had been cracked through the center and refrosted in a sickly yellowish crystalline crust she recognized immediately.

Sulfur corruption.

She broke the seal and read.

The message was brief and devastating. The Verdant Fields of Ambreth — the great agricultural plains that fed nearly a third of MarBryn's population — were dying. Not from drought, not from blight, but from deliberate elemental poisoning. The nitrogen in the soil was being systematically stripped, replaced with sulfurous compounds that rendered the earth toxic. Crops were blackening. Communities were starving. And at the center of every poisoned field, investigators had found the same symbol burned into the earth: the mark of Vorrax Cindergall.

Nitra was quiet for a long moment.

"Bad news?" Brann asked carefully.

"Vorrax," she said.

Brann set down his pan. "Oh," he said. "Very bad news."

Nitra rolled the parchment carefully and looked out across the city of Verenholm, at its atmospheric processors humming steadily in the morning light, at the nitrogen-salt towers standing centuries-old and proud.

"*No Metal, No Magic,*" she thought.

And Vorrax Cindergall was stealing the metal from the land itself.

"We leave today," she said simply.

Brann picked up his pan, ate his eggs in three rapid bites, and began packing his satchels. "Obviously," he said.

Chapter 2: The Cartographer's Secret and the Road to Ambreth

The road south from Verenholm descended through the Salite Corridor — a geological wonder of MarBryn where centuries of nitrogen-rich mineral deposits had created towering formations of pale blue-white crystal that jutted from the earth like the fingers of some buried giant. Morning light fractured through them in prismatic cascades, scattering color across the road in patterns that shifted as Nitra and Brann moved through.

Nitra traveled at a measured altitude — close enough to the ground to converse with Brann, who rode a sturdy pack-beast called Churn, low enough that she could study the soil signatures below. She was already troubled. Even here, three days' travel from the Verdant Fields, she could detect a faint acidity in the air — a sulfurous undertone beneath the clean nitrogen-salt freshness of the Corridor. It was subtle. Most would never notice it. But Nitra's connection to atmospheric nitrogen was intimate and absolute, and the corruption was unmistakable.

"Tell me what you know about Vorrax," she said.

Brann adjusted his spectacles. "Vorrax Cindergall. Born in the Ashfen Wastes, eastern MarBryn. A former member of the Sulfur Alchemist Guild before they expelled him —

apparently even they thought he went too far." He reached into a satchel and produced a battered field journal. "He believes the Periodic Codex is hierarchical. That some elements are dominant, meant to rule others. His view is that sulfur — Element 16 — is the true master element. He calls nitrogen the Pretender's Gas."

Nitra's pale blue eyes narrowed almost imperceptibly. "He is wrong."

"Categorically," Brann agreed cheerfully. "But he's also extremely dangerous. He's developed a process — they're calling it Sulfur Flooding — that uses reactive sulfur compounds to chemically displace nitrogen from soil. It's not just poisoning, it's elemental theft. He's reportedly built a machine somewhere in the eastern marshes that can project the compounds over enormous distances." He paused. "A weapon, essentially. Made of metal and chemistry."

"No Metal, No Magic," Nitra murmured. "He has perverted the philosophy."

"Weaponized it, more like."

They emerged from the Salite Corridor by midday and the landscape changed dramatically. The crystalline formations gave way to rolling lowlands of amber grass and old stone-walled farmsteads. It was pleasant country. It would not remain so for long if Vorrax's campaign continued.

Near the town of Thresh-Below, they encountered trouble.

The road was blocked by four figures — not soldiers, but something stranger. They wore patchwork armor studded with yellow sulfur crystals that had been heat-fused into the metal plates, and their eyes had the flat, yellowish tinge of elemental corruption. Minor alchemical soldiers. Vorrax's foot-soldiers, chemically enhanced for aggression and dulled for obedience.

"Dragon," the foremost one said, his voice carrying a strange mechanical rasp. "Vorrax Cindergall sends his regards. Turn back."

"That seems unlikely," Nitra replied pleasantly.

Brann sighed and dismounted from Churn. "Here we go."

The soldiers moved with surprising speed, launching compressed sulfur-charge projectiles from devices strapped to their forearms — short-range alchemical launchers that fired bolts of caustic yellow energy. Nitra was already rising. She wheeled upward in a tight spiral, the first bolt passing harmlessly below her, and then she descended fast — faster than seemed possible for a creature her size — sweeping across the road with her wings generating a tremendous downdraft powered by precise nitrokinetic manipulation.

The soldiers stumbled, scattered, their footing completely destroyed by the controlled vortex she generated. One tumbled into the roadside ditch. Two more crashed into each other.

The fourth managed to fire again. This bolt caught Nitra's left wing at the outer membrane edge. The sulfur compound burned cold and bitter. She hissed — the only sound of pain she allowed — and responded with a focused blast of Cryogenesis from her throat, not flame but a concentrated column of near-absolute-cold nitrogen that flash-froze the soldier's launcher in his hand and encased his boots in a solid block of nitrogen ice.

Thirty seconds. All four soldiers were immobilized, encased to varying degrees in precisely sculpted nitrogen ice formations.

Brann looked at them with interest. "Beautiful crystalline structure, those ice formations," he observed.

"Thank you," Nitra said, landing and folding her injured wing carefully.

Brann was immediately examining the nick on her wing membrane. "Nasty. The sulfur compound is still reactive. Hold still." He produced a small flask of saline solution and a neutralizing powder from one of his satchels — baking soda-based alkaline compound, old chemistry, reliable chemistry — and cleaned the wound efficiently. "You'll be fine. Minor membrane damage."

"I know," Nitra said. "But it is noted. Vorrax knows we are coming."

"Or he suspects someone is coming. We should move faster." Brann remounted Churn. "And we should perhaps not be entirely alone."

Nitra was already thinking the same thing.

Chapter 3: Allies Found in Ambreth's Shadow

The Verdant Fields of Ambreth were dying.

Nitra saw it first from the air — enormous black patches spreading through what should have been an ocean of green and gold. The patterns were not random. They radiated outward from specific points in almost perfect geometric formations, suggesting mathematical precision in the delivery of the sulfur compounds. This was not the work of a reckless vandal. This was a carefully engineered campaign.

She landed at the edge of the nearest blackened field. The soil crumbled at her touch — desiccated, reeking faintly of sulfur dioxide. She pressed one clawed hand into the earth and closed her eyes, reaching with her elemental senses. The nitrogen was almost entirely gone. The molecular pathways through which nitrogen-fixing bacteria had worked for centuries were destroyed, replaced with sulfate chains that poisoned the microbial ecosystem entirely.

"It's worse than the message described," she said quietly.

"It always is," said a new voice.

Nitra turned. A tall woman stood at the field's edge — she looked to be in her mid-forties, lean and angular, with close-cropped silver hair and the deep brown skin of someone who had spent decades outdoors. She wore practical work clothes under a travel-worn coat and carried a long staff topped with a transparent crystal sphere in which small green sparks drifted and swirled.

"I am Seravine Moss," the woman said. "Botanomancer. Third Order of the Leaf, Ambreth Chapter. What's left of it." Her voice was steady but her eyes, green as living moss, were tight with grief. "I've been watching this field die for eleven days. You must be Nitra. The Council sent word you were coming."

"This is Brann Cendrel," Nitra said. "Alchemical cartographer."

Brann tipped his hat. "Charmed. Terrible circumstances."

Seravine nodded curtly and knelt at the field's edge. She placed her palm on the dead soil. The green sparks in her crystal sphere flared briefly and then dimmed. "Even my botanical magic can barely find a foothold here. The sulfate compounds have disrupted the fundamental elemental balance. Plants cannot access nitrogen without it, and nitrogen without its chemical pathways is just gas. Beautiful, inert gas."

"Not entirely inert in my hands," Nitra said.

Seravine looked at her with sudden sharp interest. "You can restore it?"

"I can attempt Fertility Infusion on a small scale — channel atmospheric nitrogen directly into the soil's molecular structure and attempt to re-establish the nitrogen cycle." Nitra paused. "But on the scale of damage I am seeing across these fields, I cannot do it quickly enough without addressing the source. As long as Vorrax's machine is active, any restoration I perform will simply be overwritten."

"The machine," Seravine said, and her expression darkened further. "We believe it is located in the Marshes of Crev — three days east. Moving it apparently. We got close once. Lost two of our Order members to his sulfur-soldiers."

"Were they killed?" Nitra asked carefully.

"No. Taken. That may be worse." Seravine stood. "I want to come with you."

"I was hoping you would say that," Nitra replied. "We need your knowledge of the land. Can you travel?"

"I packed this morning," Seravine said simply.

While Brann mapped the blackened field patterns in his journal with quick, precise notation — muttering about geometric dispersal ratios and likely projection distances — Nitra walked further into the dead field alone. She knelt and pressed both forepaws into the earth, closing her eyes, and activated Fertility Infusion.

It was delicate work. She could feel the nitrogen in the atmosphere above her, an invisible ocean of it — seventy-eight percent of every breath drawn in MarBryn — and she drew it downward through herself, channeling it into the soil in carefully modulated pulses. The biochemical pathways were damaged, not destroyed. She found the surviving nitrogen-fixing bacteria — tiny, ancient, resilient — and fed them. Around her paws, a small circle of the blackened soil shifted. Brown became dark. Dark became almost black with healthy moisture. A single blade of grass, then two, then a small cluster pushed through the surface.

Small. Painfully small against the vast devastation.

But it was possible.

Nitra opened her eyes. "Then we go east," she said.

Chapter 4: The Marshes of Crev and the Forgewright

The Marshes of Crev smelled of sulfur and stagnant water and the particular metallic tang of active alchemical processing. The four of them — Nitra, Brann, Seravine, and Churn the long-suffering pack-beast — moved carefully along the marsh's western edge, using the dense stands of ghost-reed as concealment.

Brann had produced from his largest satchel a rolled document that he spread across a fallen log: a living elemental map of the region, its surface crawling with color-coded signatures. Yellow-orange bloomed heavily across the eastern marshes. Blue-green — nitrogen markers — were almost absent there.

"The concentration suggests a stationary processing center," he said quietly. "Or a recently stationary one. It's moved before — see these trailing signatures? Like drag marks in the elemental record." He traced them with an ink-stained finger.

"It moved three days ago," a voice said from directly behind them.

Nitra spun, nitrogen already crystallizing in the air around her paws.

The figure held both hands up immediately. He was a heavyset man, broad through the shoulders, with a close-cut dark beard shot through with gray and hands that were the most extraordinary Nitra had seen on a human — each finger bore a different elemental ring, bands of various metals set with corresponding element-stones, and his forearms were covered in alchemical notation tattooed in deep indigo ink.

"I am Gordin Thal," he said carefully, eyes fixed on Nitra's frozen paws. "Forgewright. I build things. Chemical-mechanical things. I was building things for Vorrax until four days ago when I saw what his machine was actually doing to those fields." He lowered his hands slowly. "I left. He doesn't know yet. Or he didn't, four days ago."

"Why should we trust you?" Seravine asked flatly.

"Because I know where the machine is now, I know exactly how it works, and I know how to destroy it." Gordin held her gaze steadily. "Also because I'm the one who built it, and I've spent four days sick with guilt about it."

A long silence stretched across the marsh.

"Tell us," Nitra said. She had released the nitrogen ice. Her voice was calm but her pale blue eyes studied him with clinical precision. People were like elemental compounds to her — their individual components, their interactions, the stability of their bonds.

Gordin was, she assessed, genuinely remorseful. That was not nothing.

He explained. The Sulfur Flood Machine — he called it the Stripper, which he clearly hated — operated by converting atmospheric sulfur dioxide, drawn from processed fossil sulfur deposits, into reactive sulfate ions that were then electrostatically projected across distances of up to forty kilometers. It ran on a complex alchemical battery array — banks of carbon-zinc cells arranged in series — powering an enormous magnetic dispersal coil made from copper and iron. Every component was elemental. Every component was expensive. Every component had a weakness.

Brann was scribbling furiously.

"The battery array," Gordin said, leaning over the map. "Here. It's the machine's heart. Destroy those cells, the magnetic coil collapses, the projection system fails completely. The machine becomes inert." He paused. "But the cells are housed in a compound Vorrax built specifically to guard them. He's not stupid."

"What sort of compound?" Nitra asked.

"About forty sulfur-soldiers. Three alchemical turrets. A personal guard of six Cinderfists — elite soldiers, sulfur-enhanced beyond the standard. And Vorrax himself, who is..." Gordin paused, seeming to choose his words. "Powerful. He can project sulfur compounds directly without any device. It's his natural element, the same as nitrogen is yours. He knows you're coming. He's been expecting you specifically."

"Of course he has," Brann said without looking up. "Do you think the marsh has a back entrance?"

Gordin looked at him. "As a matter of fact, yes."

Chapter 5: Smoke and Crystals

They found the fifth member of their company in a small marsh village called Brinevere, its wooden structures raised on stilts above the waterline, its people pale and quiet with the anxiety of people living too close to danger.

Her name was Tessaly Wren, and she was seventeen years old and absolutely furious about everything happening to her home.

She was a small, fierce girl with close-cropped dark hair, quick dark eyes, and a talent for explosive alchemical compounds that was, Gordin said privately to Nitra, genuinely alarming for someone her age. She had no magical ability in the traditional elemental sense — but she had grown up in a family of marshland chemists who had been creating reactive compounds for fishing, land-clearing, and pest control for three generations. She knew the Marshes of Crev better than anyone alive, and she was fiercely motivated.

"Three of my family's fields were hit last week," she said, standing before Nitra with her chin up and her jaw set. "My grandmother has nothing to harvest. I want to help."

Nitra regarded her. "This will be dangerous."

"I know."

"You are young."

"I am useful," Tessaly said precisely. "There is a difference." She held up a small ceramic vial containing a faintly luminous compound. "Phosphorus-sulfate igniter. Burns hotter and faster than anything Vorrax's soldiers carry. I have fourteen of them." She paused. "Also I know a path through the eastern bog that the soldiers don't know about. The marsh gases there are explosive if you don't know the route. I do."

"She's right that she's useful," Seravine said with a faint smile. "And honestly, I've been underestimated my whole life and I turned out fine."

"That's debatable," Brann said pleasantly, and Seravine gave him a look that could have frozen nitrogen without Nitra's help.

"You come with us," Nitra told Tessaly. "But you follow guidance in direct combat situations. Agreed?"

Tessaly nodded once, sharply, already shouldering her pack.

Before the reconnaissance circuit, before the plan, there was dinner — and on the last ordinary evening any of them would have for some time, nobody skipped it. Tessaly built the fire and produced a blackened pot of her grandmother's smoked marsh-fish stew, thick with pepper-root and so sharp with spice that

Brann's eyes watered on the first bite and he ate three bowls anyway. Gordin cut dense travel bread into equal shares without being asked, the same way he did everything, precisely and without fuss, and passed around hard cheese and pickled marsh onions from his own pack. Seravine spread mossberry preserves on the last of her seed bread and shared her foraged tea, insisting — correctly — that it settled the stomach before a fight. Nitra took only her Frostberry tea, steeped dark over a small flame Brann built for exactly that purpose, and watched her companions eat with the quiet, careful attention she gave to anything that mattered to her.

"Eat," Seravine told Tessaly, when the girl tried to skip her own portion to double-check her igniters instead. "You fight better with a full stomach. So do the rest of us, watching you."

Tessaly ate.

That night, camped in the ghost-reeds with luminescent marsh-flies drifting overhead and the distant hum of Vorrax's machine audible as a low-frequency vibration through the waterlogged ground, Nitra flew a reconnaissance circuit alone.

From altitude, the compound was visible — a brutalist cluster of metal and chemical-stone structures at the marsh's eastern edge where the land firmed slightly into a raised promontory. Lights burned from multiple structures. She could see the Stripper itself from above: an enormous barrel-shaped device mounted on a rotating platform, its dispersal coil glowing faintly orange with sulfurous energy. The battery arrays were housed in a long rectangular building to the device's south.

She also saw what Gordin had not mentioned — or perhaps had not known. Vorrax had deployed a perimeter of alchemical sensor posts, crystalline devices that would detect elemental signatures. Specifically, she noted with cold recognition, tuned to detect nitrogen concentration spikes.

He had built a detection system specifically calibrated to find her.

She descended quickly below the detection threshold, angling back toward camp on a low-altitude glide over the dark water.

"*He is prepared for me,*" she thought. "*Good. That means I must be more prepared for him.*"

She reviewed the compound's layout in precise mental detail as she flew, her analytical mind mapping every structure, every guard position, every structural weakness.

By the time she landed at camp, she had a plan.

Chapter 6: The Battle of the Crev Compound

Dawn came gray and mist-heavy over the Marshes of Crev.

Nitra's plan was surgical. Gordin and Brann would approach from the north using the maintenance access Gordin knew — two nondescript figures, unremarkable, goal: reach the battery building. Seravine would take the western approach through the ghost-reed beds and provide botanical interference — rapid overgrowth of the surrounding structures, buying time and confusion. Tessaly would lead Nitra through the explosive bog path and create diversionary detonations on the eastern perimeter.

Nitra herself would come from directly above.

"He expects me to announce myself with a nitrogen signature," she said during the pre-dawn briefing. "So I will not. I will approach in Mist Form, which presents minimal elemental signature, and I will wait for your diversions before I act."

"And then?" Tessaly asked.

"And then I will act very decisively."

"I love it when she says things like that," Brann said. "Cheerfully ominous."

The assault began at first light.

Tessaly moved through the bog path with the sure-footed confidence of someone who had walked it a hundred times, Nitra drifting alongside her in Mist Form — barely visible, a faint blue-white haze above the waterline. Three phosphorus-sulfate igniters, precisely placed at Tessaly's direction, detonated along the eastern perimeter in rapid sequence. The explosions were not enormous but they were spectacular — hot white phosphorus light and concussive force that shattered two sensor posts and sent six sulfur-soldiers scrambling.

Simultaneously, the north wall of the compound suddenly erupted in dense vegetation — Seravine's botanomancy working through years of dormant marsh-plant roots, driving them upward through the compound's foundation, cracking the base of the guard wall, forcing soldiers off their positions.

In the confusion, Gordin and Brann slipped through the maintenance gate. Gordin moved with practiced familiarity through corridors he had designed himself. Brann was three steps behind, already identifying structural load points with professional calm.

And Nitra materialized above the compound's center, reconstituting from Mist Form at altitude — fully solid, wings spread enormously, catching the gray dawn light and turning it into something blazing and blue-white.

Every face in the compound turned upward.

"Nitrogen dragon," one soldier breathed.

"Correct," Nitra said, and exhaled.

The Cryogenesis wave swept across the compound in a controlled arc — she was precise, always precise — targeting the alchemical turrets specifically. All three turrets flash-froze solid, their sulfate projection mechanisms locked immovably in nitrogen ice. Three turrets silent in less than four seconds.

The Cinderfists came next. They launched themselves at her with sulfur-enhanced speed, their armor crackling with yellow elemental energy, and they were fast. Genuinely fast. Two of them caught her mid-wing with sulfate bolts that burned through membrane tissue, and she felt the hot-cold pain of the chemical reaction clearly.

She responded with a technique she privately called the Vortex Lock — a concentrated nitrokinetic vortex, pulling nitrogen from the atmosphere in a swirling column that caught all six Cinderfists simultaneously and drove them into the compound's eastern wall with enough force to knock them senseless.

She was bleeding from two wing wounds. She did not stop.

Inside, Gordin was working rapidly on the battery array with tools he had brought specifically for this. "Third series, bypass the safety coupling—" he muttered, hands moving with practiced expertise. Brann handed him components as directed, reading the array architecture with sharp eyes. A shot from a sulfur-soldier caught Brann across the shoulder, knocking him hard into the battery housing. He slumped, gasping, face white with shock and pain.

"Brann!" Gordin grabbed him.

"I'm fine," Brann said unconvincingly. "Keep working. Keep — working." He pressed his hand against his shoulder, breathing carefully, and with his other hand kept his torch steady so Gordin could see.

Seravine fought her way through the north wall breach, using her botanomancy to drive root systems through the floor at soldiers' feet, entangling them with aggressive efficiency. Tessaly, outside, was running along the perimeter placing her remaining igniters at structural points Nitra had identified.

And then Vorrax Cindergall stepped from the main building.

Chapter 7: Vorrax Speaks

He was not what Nitra had expected.

Vorrax Cindergall was lean and almost scholarly in appearance — long-fingered, precisely groomed, wearing a coat of deep amber over which sulfur-crystal formations grew organically, as if he were both wearing and becoming his element. His eyes were the color of old sulfur: a cloudy, yellowish brown that held cold intelligence.

He looked up at Nitra hovering overhead and he smiled.

"Element Seven," he said. His voice carried remarkable calm. "I wondered when the Council would send their dragon."

Nitra descended slowly, landing twenty meters from him in the compound's central yard. She kept her wings partially extended — ready to rise instantly, or to generate a vortex. Her wounds throbbed but her expression was composed.

"You are poisoning the Verdant Fields," she said. "Three hundred thousand people depend on those fields, Vorrax."

"Three hundred thousand people," he repeated thoughtfully. "Yes. And in ten years, six hundred thousand. And then a million. MarBryn cannot sustain exponential growth on finite elemental resources, dragon. I am not poisoning the Fields. I am enforcing a correction."

"You are enforcing your own dominion," Nitra said. "Under the pretext of philosophy."

His eyes sharpened. "And you are enforcing yours. The difference is, I am honest about it." He raised one hand, and yellow-orange sulfur energy crackled between his fingers with casual power. "Nitrogen and sulfur have always competed in natural systems. In soils, in atmospheres. I am simply accelerating what nature already intends."

"Nature does not intend starvation," Nitra replied. "The nitrogen cycle exists specifically to prevent it. You are not accelerating nature. You are breaking it."

Vorrax tilted his head. "We shall see whose element is stronger today."

He moved very fast.

The sulfur projection that hit Nitra was nothing like his soldiers' weapons — it was a concentrated burst of SO_3, sulfur trioxide, which reacted violently with atmospheric moisture to form sulfuric acid vapor on contact. It struck her chest-scales with searing force, and even her dense cobalt scales felt it — a fierce burning, chemical rather than thermal.

She absorbed the hit and immediately responded with a wave of Atmospheric Nitrate Manipulation — flooding the immediate area with concentrated nitrate compounds that reacted with his sulfuric vapor, neutralizing it in a cascading chemical chain. White crystalline residue showered the compound yard between them.

Vorrax blinked. He had not expected that.

"Neutralization chemistry," she said calmly, wings spreading fully. "No Metal, No Magic, Vorrax. But chemistry has always been about balance."

He launched three more projections in rapid sequence. She met the first with Cryogenesis, freezing the sulfur trioxide before it could react with moisture. She evaded the second with a sharp nitrokinetic-assisted aerial pivot that took her ten meters sideways in less than a second. The third she caught on a nitrogen ice barrier she erected between them — a half-meter thick wall of solid nitrogen crystal that the projection simply could not penetrate.

Through the wall, she could see him clearly. And she could see that, for the first time, his composure had cracked slightly.

Inside the battery building, Gordin completed his sabotage. He disconnected the seventh and final battery series coupling, bypassed the safety arrestors, and triggered a controlled chemical cascade through the array — not explosive, but complete. The batteries discharged harmlessly into the ground through a shunt he had spent an hour installing.

The Stripper's coil went dark.

The machine let out a long, descending mechanical moan and fell silent.

Even outside, across the compound, they all heard it.

Vorrax heard it too. His expression shifted — the cool intelligence replaced, briefly, by raw fury.

"The machine is dead, Vorrax," Nitra said. "Stop this."

For a moment he stood completely still. Then, controlled again, lethal again, he turned and ran.

Chapter 8: The Cost of the Chase

Nitra went after him immediately.

She transformed partially into Mist Form as she rose, allowing her to move through the compound's structures rather than around them, reconstituting on the far side to see Vorrax already at the marsh's edge, running with the supernaturally enhanced speed of a sulfur-elemental. He was fast. On open ground, even she might struggle to close the distance.

She dove.

Her wings generated what she thought of as a Gale Turn — a massive nitrokinetic surge, tucking and then snapping open at low altitude to create a directed pressure wave that hit the marsh surface ahead of Vorrax, geysering dark water and forcing him to veer sharply left.

Into the eastern bog path.

Nitra pulled up hard. "*The explosive marsh gases,*" she thought. "*Tessaly's path. But he doesn't know it.*"

She circled at safe altitude and watched. Vorrax's enhanced speed was carrying him deep into the bog before his alchemical senses presumably detected the methane and hydrogen sulfide concentrations rising around him. She saw him slow, uncertain.

Tessaly appeared at the bog's edge, breathing hard from running. She looked up at Nitra circling above.

"Can you freeze the gas pockets?" she shouted. "Solidify the surface above them without igniting anything?"

Nitra assessed in seconds. Cryogenesis, precise temperature control, targeted freezing of the surface methane without triggering combustion. Difficult. Not impossible.

"Get back," she called down.

She descended to the lowest safe altitude and began. It required extraordinary precision — lowering the temperature of specific gas pockets rapidly enough to prevent any ignition chain, solidifying them into harmless frozen crystite formations. The blue-white cold radiated from her in controlled pulses, and she moved systematically across the bog ahead of Vorrax's position, cutting off his escape routes one section at a time.

Vorrax tried to project sulfur energy to clear a path. The sulfur projections, meeting the frozen bog surface, produced unpredictable chemical reactions — sulfur dioxide in a methane-hydrogen sulfide environment, even partially frozen. One small detonation knocked him from his feet. He rose, limping.

He was caught.

But he was not finished. He turned and directed a full-power projection upward at Nitra — not sulfur trioxide this time, but pure concentrated sulfur dioxide, an asphyxiant, aimed at her breathing passages. It caught her fully in the face at close range.

Nitra faltered in the air. Her Mist Form reflexes activated partially — an instinctive defense — and the sulfur dioxide passed partly through her mist-state, but not entirely. She coughed, struggled, dropped ten meters before regaining controlled flight. The chemical burn in her upper respiratory passages was severe.

She was hurt. Genuinely hurt.

She pushed through it with cold discipline and unleashed a Deep Freeze — the largest sustained Cryogenesis field she had ever attempted, not targeted but broad, dropping the temperature across the entire bog section by forty degrees in under ten seconds. The bog surface froze solid. Vorrax was encased to the knees in nitrogen ice, his movement completely arrested.

Nitra landed beside him, breathing carefully around the burning in her throat.

"It is over," she said Her voice was raw but steady.

Vorrax looked up at her from his frozen prison. The fury in his yellowed eyes had collapsed into something colder and more complicated — not defeat exactly, but the expression of a man recalculating every variable simultaneously.

"You think destroying one machine ends this?" he said.

"No," Nitra replied honestly. "But it ends today's chapter of it."

Back at the compound, the situation was being sorted with the methodical efficiency of people who had survived something terrible and were now dealing with the aftermath. Gordin had bound Vorrax's sulfur-soldiers with chemical neutralizing compounds — binding agents that suppressed elemental enhancement — working alongside Seravine, whose botanomancy provided additional restraint in the form of rapidly grown marsh-vine shackles. It was, Brann observed from where he sat against the battery building wall with his shoulder bandaged in strips torn from Gordin's spare coat, a remarkably effective combination of old botanical magic and modern alchemical chemistry.

"No Metal, No Magic," Brann said, somewhat deliriously from pain and blood loss. "Plants count as organic chemistry. Did you know nitrogen is in chlorophyll? Nitra is literally in the plants."

"Stop talking," Seravine told him, not unkindly, tightening his bandage with practiced hands. She had field medical training alongside her botanomancy — the two disciplines, she had always found, went naturally together. Healing plants and healing people shared more methodology than most people realized.

"My shoulder really hurts," Brann confided.

"I know. You were shot with a sulfate bolt. It will hurt considerably more tomorrow."

"Wonderful. Something to look forward to."

Tessaly, who had run the perimeter to check for remaining soldiers and found none — most had fled after the machine fell silent — dropped down beside Brann and handed him a flask of clean water. She looked at his shoulder with the assessing eyes of someone raised around chemical injuries.

"Clean wound," she said. "The bolt didn't fragment. You'll keep the shoulder."

"Tremendous news," Brann said, and passed out briefly.

Chapter 9: The Healing of Ambreth

They held Vorrax Cindergall for two days in the Crev compound while Gordin disassembled the Stripper with meticulous care, cataloguing every component for the Nitrate Council's records. He worked with the focused, quiet dedication of a man performing penance, and Nitra respected that. She did not hurry him.

On the second day, a Council transport arrived — a large chemical-powered dirigible, its envelope filled with nitrogen gas, its hull plated with the pale nitrogen-stone of Verenholm's towers. Eight Council marshals in blue-gray uniforms took custody of Vorrax and his soldiers without ceremony. The Council's chief arbiter, a stern woman named Orrath Pell, shook Nitra's forepaw with both hands and said nothing for a long moment.

"The fields?" Nitra asked.

"Damaged. Severely. But your stopping the machine when you did—" Orrath paused. "The projections had been running for eleven days. Another eleven and the damage would have been irreversible across ninety percent of the cultivated area. You stopped it at sixty percent damage. Which means forty percent is intact." She looked out across the marsh toward the distant west. "It is still a catastrophe. But it is a survivable one."

Nitra looked at her companions. Brann, his arm in a proper sling now, managed a tired smile. Seravine stood straight-backed and dry-eyed but her green eyes were bright. Tessaly was watching the Council transport with an expression of fierce satisfaction. Gordin had his back to all of them, still working on the machine's dismantlement, not wanting to see Vorrax taken away — guilt made strange architects of people's attention.

"Then we go to Ambreth," Nitra said.

The journey west took four days, traveling with Brann on Churn and the others on horses provided by the Council. Nitra flew the route each day and returned to camp each evening, using her time in the air productively. She was mapping the atmospheric nitrogen concentrations across the region — carrying Brann's portable elemental sensors strapped to her chest in a specially designed harness Gordin had improvised — building a precise picture of where the soil damage was worst, where the nitrogen cycle was most disrupted, and where restoration should begin.

Her wing wounds had been properly dressed by the Council's field medic and were healing with the resilience of her elemental nature. Her respiratory passages, burned by Vorrax's sulfur dioxide attack, were improving daily. She still felt it when she flew at altitude, a rawness in each breath, but she flew anyway. The work was more important than her discomfort.

On the third night of travel, Gordin sat beside her at the campfire. He was quiet for a long time, turning one of his elemental rings over and over on his finger — the nitrogen-stone ring, pale blue, which he had worn longest of all.

"I want to help restore the fields," he said finally. "Not as — not as absolution. I know it doesn't work like that. But because I know the chemistry of what the Stripper did better than anyone, which means I know the reversal chemistry better than anyone too."

Nitra was quiet, considering. "The reversal chemistry will require both elemental magic and mechanical application," she said. "Atmospheric nitrate infusion from me to re-establish the nitrogen baseline. Mechanical tilling and chemical amendment to support the microbial recovery. And botanical encouragement to accelerate the plant-root establishment."

"That's exactly what I was thinking," Gordin said. He finally stopped turning the ring. "A three-system approach. Magic, chemistry, botany. Working simultaneously."

"It would take months," Seravine said from across the fire. She was not objecting. She was assessing.

"Yes," Nitra agreed. "It would."

"Then we had better start," Seravine said.

Tessaly looked up from where she was carefully inventorying her remaining compounds. "I'm staying," she said simply. "My family's fields are in the damaged zone. I'm not going anywhere."

Brann, propped comfortably against his saddlebag with his arm in its sling, had his journal open across his knee. "I have mapped every elemental signature across the damaged area," he said. "I can produce prioritized restoration zone maps that tell you precisely where to start for maximum ecological impact. Working outward from the healthiest surviving nitrogen pockets." He looked at Nitra. "I would also like to stay. The cartographic documentation of an elemental restoration of this scale would be—" he searched for the word, "unprecedented."

"Also you would not leave me here alone with all of you serious people," Nitra said, with the faintest warmth in her composed voice.

Brann grinned. "That too."

They reached the Verdant Fields of Ambreth on the fourth day, and Nitra flew the full extent of the damaged zone before landing. It was heartbreaking from the air — a vast patchwork of black and brown amid the surviving green, like a map of war drawn on living earth. Farmers stood at field edges looking at nothing.

Children who should have been running between crop rows sat quietly on stone walls.

But there were also people already working. Neighbors carrying water. Community chemists setting up small-scale soil amendment stations with whatever materials they had. An elderly woman — Tessaly's grandmother, it turned out, a tiny fierce person named Corra who shook Nitra's forepaw without any apparent intimidation whatsoever — was directing a group of volunteers spreading alkaline amendment powder across a blackened stretch.

"It's not enough," Corra said bluntly. "But it's something. Doing something is always better than doing nothing."

"That," Nitra said respectfully, "is exactly right."

She began that afternoon.

Standing at the center of the most severely damaged field, Nitra opened her connection to the atmospheric nitrogen above — that vast invisible reservoir, seventy-eight percent of every breath, the quiet backbone of all life in MarBryn — and began to draw it down. Not in pulses this time, not in the careful small doses she had managed before, but in a sustained, powerful, precisely modulated flow that she had never attempted at this scale.

It was exhausting in a way she had not anticipated. The elemental magic was not the difficulty — she had more than enough power for the task. The difficulty was the precision required. Too much nitrogen delivered too fast would oversaturate the soil chemistry and create its own toxicity. She had to feel each section of earth beneath her, assess its current chemical state through her elemental senses, and calibrate the infusion to exactly what each microscopic zone of soil could accept and process.

She worked for six hours on the first day without stopping.

When she finally raised her head, the field immediately around her — perhaps half a hectare — had shifted from dead black to dark, rich brown. It was not green yet. Green would take weeks. But the nitrogen was there, restoring the microbial pathways, laying the molecular groundwork for recovery. She could feel the surviving bacteria in the soil responding, tentative at first and then with gathering biochemical enthusiasm.

Seravine worked alongside her, her botanomancy coaxing dormant root systems and seed banks that had survived in the soil's deeper layers. Gordin supervised teams of local volunteers applying precisely formulated chemical amendment mixtures he designed on-site using materials sourced from Brann's comprehensive supply list. Tessaly ran between stations, mixing compounds, answering questions, showing local chemists techniques her family had used for generations in the marsh.

And Brann mapped everything, his living elemental charts updating in real time, his annotations growing more detailed and more optimistic with each passing day.

At the end of the first week, three hectares had been restored.

At the end of the second week, eleven.

Chapter 10: What Grows After Winter

Two months later, the Verdant Fields of Ambreth were green.

Not entirely — the most severely damaged zones would need another season of careful remediation, and Gordin had established a permanent chemical amendment station that would operate for at least two years under the supervision of local alchemical chemists he had trained. But the majority of the fields were alive, and alive with a vigor that surprised even Nitra's careful projections. The nitrogen-cycle restoration, combined with Seravine's botanical accelerants and Gordin's soil chemistry expertise, had produced an ecosystem recovery faster than any of the scientific literature of MarBryn had recorded.

"The surviving microbial communities were more resilient than standard models predicted," Gordin said one morning, reviewing his field data at the long wooden table outside Corra's farmhouse. "Once the nitrogen baseline was restored and the sulfate toxicity neutralized, they rebounded exponentially. The soil essentially remembered how to be healthy."

"Elements remember their purpose," Nitra said, from where she sat at the table's end with a bowl of the strong dark tea that she had developed a fondness for over the past weeks. "That is not so different from people."

Gordin was quiet for a moment. "No," he agreed. "I suppose it isn't."

He had submitted a full technical account of his involvement in building the Stripper to the Nitrate Council, voluntarily and completely. The Council's arbiter, Orrath Pell, had communicated her decision: Gordin would face a period of supervised service to the Council's restoration programs — three years, working on elemental remediation projects across MarBryn. It was not a punishment he resisted.

"It's exactly what I would have chosen," he said. "If I'd been thinking clearly from the start."

Vorrax Cindergall was held in the Council's elemental containment facility in the capital city of Meridane — a modern marvel of alchemical architecture where elemental abilities could be safely suppressed without harm to the person. His trial was scheduled. His soldiers, once the sulfur-enhancement compounds had been medically cleared from their systems, had mostly turned out to be people who had been recruited under false pretenses and chemical coercion. The Council's medical division was managing their recovery carefully.

"It's not simple," Seravine said one evening, when this subject came up. "Vorrax used people the way he used the soil — extracted what he wanted and left the rest poisoned." She paused.

"That's the part that makes me angriest, I think. Not even the fields. The people."

"The fields and the people," Tessaly said firmly. "Both."

Seravine looked at her with a slight smile. "Both," she agreed.

The morning before Nitra planned to leave, she flew the full extent of the Verdant Fields one final time. She rose before dawn and climbed until Ambreth was spread below her in the gray pre-light — and then the sun found the horizon and the fields turned gold and green in the new day, and the atmospheric nitrogen above them carried the clean, rich scent of living soil and growing things.

She hovered in the current, wings barely moving, held aloft by the same element she had spent months pouring into the earth below. The healing of Ambreth had cost her more than she had expected — in physical effort, in elemental expenditure, in the complicated feelings that came with having faced an adversary who believed, genuinely believed, that he was right.

Vorrax was not stupid. He was not simple. His philosophy was wrong — wrong in its arrogance, wrong in its willingness to destroy lives for the sake of a theoretical correction — but it had emerged from a real observation about a real problem. MarBryn was growing. Resources were finite. These were true things. The answer was not to weaponize chemistry against the vulnerable. The answer was what she and her companions had demonstrated over two months of daily work: that understanding the elements, working with their natural cycles rather than against them, produced outcomes that sustained rather than destroyed.

"*No Metal, No Magic*," she thought, riding the high current above the living fields. "*And no life without nitrogen*."

She descended.

Her companions were gathered outside Corra's farmhouse when she landed — all five of them, which surprised her slightly. She had assumed some goodbyes would happen quietly, individually.

Brann had his arm out of its sling now, though he still favored it. He had produced, from the depths of his largest satchel, a final completed map — a full elemental survey of the Verdant Fields of Ambreth, pre-damage and current status overlaid in two colors, the restoration progress annotated in his meticulous handwriting. He presented it to Nitra with both hands.

"For the Council's records," he said. "And for yours."

Nitra accepted it with care. The map was extraordinary — a living document of what had happened here, what had been done, what had been saved. She rolled it carefully and secured it in the travel case that hung at her side.

"Thank you, Brann," she said. "For all of it."

He shrugged his good shoulder. "I was going to map something. Might as well have been something that mattered." He paused. "I've been offered a position with the Council's elemental survey division, actually. Based in Meridane. Permanent post, proper equipment." He looked slightly dazed by this, as if the respectability of it had not quite settled yet. "I think I'm going to take it."

"You should," Nitra said. "MarBryn needs people who can read its elemental story."

Seravine was returning to what remained of the Ambreth Chapter of the Order of the Leaf — rebuilding it, she said, since it needed rebuilding and she was apparently the person to do it. She had also, over the past two months, begun teaching Tessaly the fundamentals of botanomancy alongside the girl's native chemical knowledge. Tessaly had, it emerged, a genuine latent sensitivity to plant elemental signatures — not a full botanomancer's gift, but a natural affinity that Seravine said was worth cultivating.

Tessaly herself was staying in Ambreth. Her family was here. Her grandmother was here. The fields were here. She had applied, at

Gordin's and Seravine's joint urging, to the Nitrate Council's junior alchemist scholarship program — a fully funded three-year study in applied elemental chemistry based in Verenholm.

"You'd be a natural," Gordin had told her.

"I know," Tessaly had replied, with exactly the directness that had characterized her since the moment she had stood before Nitra in Brinevere with her chin up and her case made.

Gordin himself departed that same morning for Meridane, where his Council service assignment began in a week. He shook hands with each of them — a quiet, firm handshake that carried the weight of everything he had done wrong and was trying to do right. When he reached Nitra, he paused.

"I'm going to spend a long time making up for what the Stripper did," he said.

"Making things grow again is not a bad way to spend a life," Nitra told him.

He nodded once and walked toward the road north.

Nitra stood for a moment with Brann and Seravine and Tessaly in the morning light outside Corra's farmhouse while Corra herself appeared in the doorway with a tray of tea and made it immediately clear that no one was leaving until they had drunk it, which was not an instruction anyone disobeyed.

It was not only tea. Corra had been cooking since before dawn, and the long outdoor table filled quickly with a spread that read, without anyone saying so out loud, like a roll call of the people gathered around it. There were garden tomatoes sliced thick and salted the way Corra liked them best. There was a pot of smoked marsh-fish stew, spiced fiercely enough that Brann reached for water twice before he reached for a second helping. There was dense bread and hard cheese set out at Gordin's elbow, and a jar of mossberry preserves that Seravine had been quietly restocking for weeks. Brann's soft-boiled eggs sat in a small dish near his hand, because Corra had asked Tessaly what the cartographer ate every morning and had simply provided it. And at the head of the table, Nitra's cup of Frostberry tea steamed dark beside a bowl of chilled cloudmelon that Corra had put on ice specifically for her, without being asked, the same way she seemed to know everything else that needed doing before anyone requested it.

"Eat first, talk after," Corra announced, in a tone that closed the subject.

Nobody argued with her. Nobody ever did.

They sat together at the outdoor table — a dragon, a cartographer, a botanomancer, a marsh chemist, and one extremely small elderly farmer who had not been intimidated by a single event of the past two months — and they drank their tea while the fields of Ambreth breathed in the morning around them.

"Where will you go next?" Seravine asked Nitra.

Nitra considered the question genuinely. The Nitrate Council would have other assignments — there were always other problems in a world as complex and elemental as MarBryn. There were reports of nitrogen depletion in the northern highland farming communities, possibly natural, possibly not. There were ancient nitrogen-stone towers in the Salite Corridor showing structural crystal degradation that no one had explained. There was always something.

"North, I think," she said. "The highland communities. And then perhaps the Salite Corridor."

"The crystal degradation in those towers," Brann said immediately, professional interest lighting his face. "I've noted anomalous elemental signatures in two of them. Something is interfering with the nitrogen-salt matrix at the molecular level. Could be natural weathering. Could be—"

"Could be something we need to investigate," Nitra said.

"Exactly what I was going to say."

"You are coming north with me," Nitra said. It was not a question.

Brann looked at his newly offered permanent position in Meridane, at his sling, at his battered satchels, at Churn the pack-beast who was currently eating Corra's garden hedge. He looked at the map he had made of a restoration he had been part of. He looked at Nitra.

"Obviously," he said.

Corra leaned forward and refilled everyone's tea without being asked, with the authority of a woman who had been doing exactly what was needed long before anyone thought to request it. She looked at Nitra with her sharp, unsentimental eyes.

"You'll come back," she said. It was not a question either.

"To see the fields through another season," Nitra agreed. "Yes."

Corra gave a satisfied nod and settled back in her chair.

An hour later, Nitra stood at the field's edge where she had first pressed her paws into the dead earth two months before. She pressed them there again now. The soil was warm and dark and alive, teeming with the invisible biochemical industry of a nitrogen cycle restored. She could feel it — the bacteria, the root networks, the slow molecular conversations of a living ecosystem conducting the patient chemistry of growth.

"*Element 7,*" she thought. "*Nitrogen. Invisible, essential, everywhere.*"

No Metal, No Magic.

No life without the elements that bind it.

She raised her wings — those great feathered-ribbed wings that caught the morning light and scattered it in pale blue and silver — and rose from the earth in one clean, powerful motion. Below her, the Verdant Fields of Ambreth stretched green and gold and healing to the horizon, and the nitrogen-rich air carried her upward with the effortless generosity of an element that asked nothing in return for sustaining everything.

Brann mounted Churn on the road below and looked up at her, hand shading his eyes.

"North?" he called.

"North," she confirmed, already climbing.

She rose until the fields were a tapestry below her and the sky was vast and clear and full of the invisible chemistry that made all things possible — and then she banked north, wings steady, breath clean, the world of MarBryn spread beneath her in all its elemental, endlessly complicated, endlessly remarkable entirety.

The wind was good.

Nitra flew.

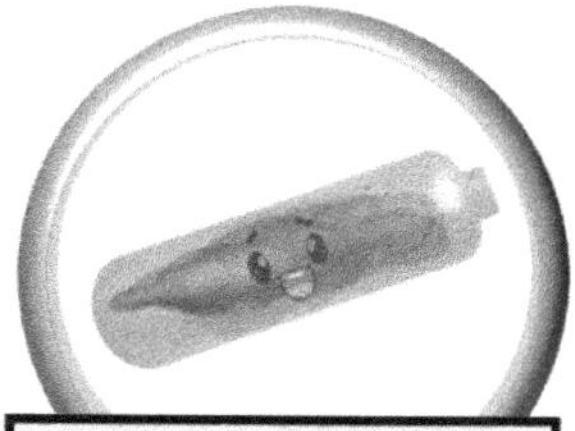

Nitrogen Facts

- Officially discovered in 1772 by Scottish physician D. Rutherford
- Colorless, odorless, tasteless, and inert gas
- Poor conductor of heat and electricity
- Diamagnetic
- Neither ductile or malleable
- Non-Metal—Gas

Author's Note

When I began imagining Nitra, the dragon with Nitrogen magic, I wanted to weave together two deeply satisfying curiosities: the beauty of magical worlds and the real-world wonder of chemistry. No Metal, No Magic is not just a philosophy for a fantasy realm; it is a reminder that every act of creation—whether in a forge, a laboratory, or a classroom—rests on understanding the elements that bind our world together. Nitrogen, the quiet giant of the atmosphere, became the heartbeat of Nitra's power: a symbol of potential, discipline, and stewardship.

Nitra's magical journey grew from a million questions: What if a dragon could teach us to observe, measure, and balance? What if the most devastating conflicts could be resolved by collaboration across disciplines—alchemy, botany, materials science, cartography, and ecological restoration? I wanted a story where science is not dry fact but a living, lyrical force that partners with magic to heal and rebuild. The result is a tale of courage, responsibility, and hope—where a chosen guardian, a round-table of unlikely allies, and a cautious, ever-curious world must reckon with the consequences of power and ambition.

Nitra's journey is as much about growth as it is about battle. Her cryogenesis and nitrokinetics are not mere weapons; they are tools for renewal. The companions who join her—Brann,

Seravine, Tessaly, Gordin—are imperfect, passionate, and undeniably brave. Their dynamics—humor, stubbornness, sacrifice—mirror the messy, wonderful reality of teamwork. Through them, I wanted to celebrate diverse ways of knowing: the cartographer who reads the land, the botanomancer who speaks to roots, the marsh chemist who tinkers with recipes for resilience, and the engineer who carries guilt and makes hard choices for the common good.

If you read Nitra and feel inspired to look at the world with a scientist's curiosity or a storyteller's wonder, I've done my job. May your days be filled with questions as bright as a nitrogen plume, and may you find in the pages of this book a reminder that true power lies in balance, care, and the courage to fight for a future that nourishes all living things.

Sybrina Durant

PS: If you're interested in reading the resources the scientific data in this book was derived from, check out this webpage:
https://magicalptelements.com/mdapt-2/

About The Author

Sybrina Durant is the author of

"The Magical Elementals Chemistry Fantasy Series"

"The Magical Elements of the Periodic Table Series"

"The No Metal No Magic Series"

"The Blue Unicorn – Journey To Osm" books

and she is the creator of the "Elemental Fantasy Periodic Table" which is available as a brilliantly colored poster.

There are also book activities, songs, trading cards and t-shirts for all 118 elements and elementals, too.

Visit MagicalPTElements.com to see everything that is currently available.

If you love to sing and learn, check out the "Chemistry Sing Alongs" playlist on the Sybrinad channel on Youtube.

Glossary of Scientific Terms in Nitra the Elemental Dragon with Nitrogen Magic

Nitrogen (Element 7): The essential non-metal gas that makes up 78% of Earth's atmosphere and, in MarBryn, is the invisible backbone of all life, plant growth, and atmospheric power. It is Nitra's own element — Every technique Nitra uses in this story has a name and a real chemistry lesson underneath it. None of it is invented from nothing — nitrogen really can freeze things, really does make up most of the air you're breathing right now, and really is what plants need to grow. Nitra's magic is that science, sped up and aimed on purpose.
Story Magic: The invisible backbone of Nitra's power — the element she embodies, breathes, and bends to protect life across MarBryn.
Real Science: A colorless, odorless non-metal gas that makes up about 78% of Earth's atmosphere. It is essential to every living thing: it's a key building block of proteins and DNA, and nearly all of it enters living systems through the nitrogen cycle.

Periodic Codex
Story Magic: MarBryn's version of the periodic table — the sacred chart Vorrax twists into a ladder of dominance, with sulfur at the top.
Real Science: The real periodic table organizes all known elements by atomic number and properties. No element is scientifically "stronger" or "weaker" than another — they simply behave differently, which is the real-world fact Vorrax's philosophy gets wrong.

Element & Atom
Story Magic: In MarBryn, every element carries its own kind of magic, and every elemental being (like Nitra) is a living expression of one.
Real Science: An element is a pure substance made of only one type of atom — the smallest unit of matter that keeps that element's identity. Nitrogen atoms, sulfur atoms, and so on are the real-world building blocks the whole Periodic Codex is organized around.

Compound & Molecule

Story Magic: Vorrax's weapons and Nitra's counters are both described as "compounds" throughout the story — chemistry given a side to fight for.
Real Science: A molecule is two or more atoms bonded together; a compound is a molecule made of two or more different elements. Nitrogen gas (N_2) is a molecule but not a compound, since it's only nitrogen atoms; sulfate (SO_4^{2}□) is both, since it combines sulfur and oxygen.

Diatomic Molecule (N_2)
Story Magic: The tough, twin-atom form nitrogen takes in MarBryn's air — stable until Nitra or nature's bacteria give it a reason to change.
Real Science: Real nitrogen gas exists as N_2, two nitrogen atoms locked together by one of the strongest bonds in chemistry (a triple bond). Breaking that bond takes real energy, which is exactly why nitrogen-fixing bacteria — and lightning — are so important in the real nitrogen cycle.

Inert Gas
Story Magic: Ordinary nitrogen is called "beautiful, inert gas" by Seravine — stable until Nitra gives it purpose and direction.
Real Science: An inert gas resists reacting with other chemicals under normal conditions. Nitrogen's triple bond (see Diatomic Molecule) makes it naturally unreactive, which is why it needs an energy source — lightning, industrial processes, or bacteria — to be pulled apart and put to use.

Diamagnetic
Story Magic: A quiet, independent trait that mirrors Nitra's own character — she does not force her way into a fight or cling to what isn't hers.
Real Science: Diamagnetic materials are weakly repelled by magnetic fields rather than attracted to them. Elemental nitrogen genuinely has this property, unlike a metal such as iron, which is strongly attracted to magnets.

Ion
Story Magic: Sulfate ions are the specific weaponized particles Vorrax's Sulfur Flooding machine injects into the soil.

Real Science: An ion is an atom or molecule that has gained or lost electrons, giving it an electrical charge. Sulfate ($SO_4^{2\square}$) is a real ion with a negative charge, and its charge is part of why it disrupts soil chemistry the way it does.

Elemental Signature
Story Magic: The chemical "fingerprint" Vorrax's sensors are tuned to detect — specifically tuned to nitrogen, which is why Nitra must approach in Mist Form.
Real Science: Every chemical substance really does leave a detectable trace — through smell, reactivity, or instruments like spectrometers that identify elements by the exact wavelengths of light they absorb or emit.

Nitrokinetics
Story Magic: Nitra's specialized power to gather and direct atmospheric nitrogen as wind, pressure waves, and vortices.
Real Science: Since nitrogen is 78% of the air, moving nitrogen at scale is simply moving wind. Real wind is driven by pressure differences — Nitra's magic is the ability to create and direct those differences precisely.

Cryogenesis
Story Magic: Nitra's cold breath — a concentrated blast of near-absolute-cold nitrogen that freezes weapons, traps enemies, and builds ice walls.
Real Science: Nitrogen becomes a liquid at about -196°C (-321°F). Liquid nitrogen is used in the real world for flash-freezing food, preserving biological samples, and even removing warts — genuinely one of the coldest common substances people work with.

Mist Form
Story Magic: Nitra loosening her body into a barely visible, low-signature blue-white haze to slip past detection.
Real Science: This is inspired by real phase changes — the way matter shifts between solid, liquid, and gas. Gas particles are naturally spread out and dispersed, which is the real property Mist Form exaggerates into stealth.

Nitrogen Ice / Crystite

Story Magic: The solid, sculpted nitrogen crystal Nitra forms to freeze enemies in place or wall off an attack.
Real Science: Solid nitrogen is real: below about -210°C (-346°F), nitrogen gas freezes into an actual crystalline solid, similar in concept to how water freezes into ice, just at a vastly colder temperature.

Atmospheric Nitrate Manipulation
Story Magic: Nitra's counterattack against Vorrax's acid — a wave of concentrated nitrate compounds that neutralizes his sulfuric acid vapor.
Real Science: This is acid-base neutralization: an acid and a base-forming compound reacting to cancel each other out. It's the same basic chemistry as an antacid tablet calming an upset stomach, scaled up to a duel between elementals.

Nitrogen Fixation
Story Magic: The natural process Vorrax's machine destroys and Nitra's Fertility Infusion restores — soil bacteria feeding the whole ecosystem.
Real Science: A real biological process: specialized bacteria convert unusable nitrogen gas (N_2) into nitrates and ammonia that plants can absorb through their roots. Without it, most plant life on Earth would starve for nitrogen despite breathing air that's mostly made of it.

Fertility Infusion
Story Magic: Nitra's technique of channeling atmospheric nitrogen directly into dead soil to wake dormant bacteria and restart the nitrogen cycle.
Real Science: This dramatizes real soil restoration: adding usable nitrogen compounds to depleted soil can jump-start dormant microbial communities, which is also the basic idea behind nitrogen fertilizer — though real farmers must be as careful about over-application as Nitra is about oversaturation.

Nitrate (NO_3□)
Story Magic: The nourishing compound Nitra channels into the earth — the opposite number to Vorrax's poisonous sulfate.

Real Science: Nitrate is a real ion (NO_3□) and one of the main forms of nitrogen that plant roots can actually absorb and use to grow. It's a normal, essential part of healthy soil chemistry in the real world.

Ammonia (NH_3)
Story Magic: One of the two usable forms of nitrogen that Fertility Infusion restores to the soil alongside nitrates.
Real Science: Ammonia is a real nitrogen-hydrogen compound (NH_3) produced by nitrogen-fixing bacteria and also manufactured industrially as a key ingredient in fertilizer — it's one of the most widely produced chemicals on Earth precisely because of how much plants need it.

Botanomancy
Story Magic: Seravine's magic — the manipulation of plant biology and root systems to accelerate growth and heal crops.
Real Science: There's no real magic here, but there is real plant biology: root systems, seed banks that survive dormant in soil for years, and the biochemistry of photosynthesis are all genuine science that botanomancy dramatizes and speeds up.

Alchemical Cartography
Story Magic: Brann's skill — reading mineral and chemical signatures in soil or air to build living, color-coded maps.
Real Science: Real scientists do something similar with soil sampling and remote sensing, using chemical tests and satellite imaging to map nutrient concentrations, contamination, and vegetation health across large areas of land.

Sulfur (Element 16)
Story Magic: Vorrax's own element, positioned by the story as nitrogen's opposite — used to strip nutrients and poison soil.
Real Science: Sulfur is a real, reactive non-metal element, essential in small amounts (it's part of some amino acids) but genuinely harmful to soil and water in the large, concentrated doses the story describes.

Sulfur Flooding

Story Magic: Vorrax's destructive process for injecting reactive sulfate ions into farmland to chemically displace nitrogen.
Real Science: This mirrors a real phenomenon called soil acidification, where excess sulfur compounds (often from industrial pollution or acid rain) genuinely can lower soil pH and interfere with the nutrients plants need.

Sulfate (SO_4^{2}□)
Story Magic: The toxic ion Vorrax's machine forces into the ground, rendering it acidic and lethal to crops.
Real Science: A real compound formed from sulfur and oxygen. Some sulfate occurs naturally and is harmless in small amounts, but high concentrations really do make soil acidic and toxic to most crops.

Sulfur Trioxide (SO_3)
Story Magic: The dense gas Vorrax fires directly at Nitra, reacting instantly with moisture to form corrosive acid vapor.
Real Science: SO_3 is a real, highly reactive gas. It genuinely reacts with water vapor in the air to form sulfuric acid (H_2SO_4) almost instantly, which is part of the real chemistry behind acid rain.

Sulfur Dioxide (SO_2)
Story Magic: The choking gas Vorrax uses as a pure asphyxiant in the final confrontation, burning Nitra's respiratory passages.
Real Science: SO_2 is a real industrial and volcanic gas that is genuinely toxic to breathe in concentrated amounts, irritating the lungs and airways — which is why it is a regulated air pollutant in the real world.

Asphyxiant
Story Magic: The category of attack Vorrax uses against Nitra directly — a gas meant to interfere with breathing rather than burn.
Real Science: A real term for any substance that interferes with the body's ability to get or use oxygen. Some asphyxiant gases work by displacing oxygen; others, like concentrated SO_2, work by directly irritating and damaging the airway.

Acid & Base
Story Magic: The two chemical "sides" of the story's central conflict — Vorrax's corrosive acids against Nitra's neutralizing, base-forming compounds.
Real Science: Acids and bases are real chemical categories defined by how they behave in water — acids release hydrogen ions, bases accept them. When they react together, they neutralize each other, producing water and a salt.

pH Scale
Story Magic: The invisible scale the whole soil war is fought over — Sulfur Flooding pushes it one way, Fertility Infusion pulls it back.
Real Science: A real 0–14 scale measuring how acidic or basic a substance is (7 is neutral). Healthy farm soil usually sits close to neutral, and even small pH swings from added sulfate or nitrate compounds can make a real difference to what crops can survive there.

Neutralization
Story Magic: The reaction Nitra performs at battlefield scale, and the same basic chemistry Brann uses with a small flask on her wing wound.
Real Science: A real chemical reaction where an acid and a base combine and cancel each other out, producing a neutral, stable result — the same principle behind using baking soda to treat an acid burn or an antacid to calm heartburn.

Alkaline Compound
Story Magic: Brann's baking-soda-based treatment for Nitra's sulfur wound — old chemistry, reliable chemistry, as he calls it.
Real Science: An alkaline (basic) compound is genuinely effective at neutralizing acids. Baking soda (sodium bicarbonate) is a mild, real-world alkaline commonly used for exactly this kind of first-aid neutralization.

Phosphorus
Story Magic: The element behind Tessaly's igniters, in the same elemental family as nitrogen on the Codex.

Real Science: Phosphorus is a real element directly below nitrogen on the periodic table, sharing some of its chemical family traits. White phosphorus in particular is famous for burning extremely hot and fast, which is the real property behind Tessaly's igniters.

Methane & Hydrogen Sulfide
Story Magic: The explosive bog gases Nitra must freeze with extreme precision to trap Vorrax without triggering a blast.
Real Science: Both are real gases that build up naturally in marshes and bogs as organic matter decays. Methane is flammable and hydrogen sulfide is both flammable and toxic, which is why real bog gas is genuinely dangerous to disturb carelessly.

Biochemical Pathway / Microbial Ecosystem
Story Magic: The molecular "conversations" Nitra can feel happening in restored soil — bacteria, roots, and nutrients working together.
Real Science: A real term for the network of chemical reactions living organisms use to survive, and the real communities of soil microbes (bacteria, fungi, and more) that make that network possible. Healthy soil is genuinely full of this kind of invisible biological activity.

Nitrogen Cycle & Chlorophyll
Story Magic: The reason the fields finally turn green — nitrogen moving from Nitra's magic into the soil, into the roots, and into every leaf.
Real Science: The nitrogen cycle is the real path nitrogen takes through air, soil, and living things. Chlorophyll, the molecule plants use to turn sunlight into food, genuinely contains nitrogen atoms at its core — so restoring the nitrogen cycle really is what lets a field turn green again.